The Atkins Recipes

Boost your Metabolism with Easy and Healthy Recipes

George H. Robinson

TABLE OF CONTENTS

Introduction

You've tried losing weight but have failed. You hate to count calories or to feel that you have to restrict how much you eat, so you are always hungry. You are beginning to think that there is no hope for you.

Don't give up quite yet, because there is hope on the Atkins diet. Many people have found the answer to their weight loss issue when they start to follow the low-carb plan advocated in the Atkins diet. Instead of counting calories, you count the Net Carbs that you eat. You don't have to feel deprived, and you are encouraged to eat enough food so that you feel satiated and full.

This book will describe the Atkins diet in detail, including the four phases of the diet, the foods that you can eat on each phase, and even some recipes that you can try. All your questions will be answered in this book. By the time you are finished, you will have all the information and the tools that you need to succeed on the Atkins diet. You can lose weight fast. In fact, many people report being able to lose up to 30 pounds in 30 days!

The Atkins Diet was developed by Dr. Robert Atkins, a cardiologist, in the 1970s. However, the diet did not really catch on immediately. It wasn't until 2002, when he released a revised version of the book, Dr. Atkins New Diet Revolution, that people began to pay attention. So many people who didn't have success with typical diets which require restricting calories found that they were able to lose the weight they so desperately wanted to by doing Akins.

But what is the Atkins Diet? Basically, Atkins found that when you cut out most carbohydrates and sugars but didn't restrict other foods, people lost weight. It's not about completely cutting out carbs but includes eating only those carbs that are nutrient-dense. His diet is based on the premise that people who cut carbs and sugars but are able to eat anything else will lose weight. And the results have shown this to be true. People who start on the Atkins Diet have lost up to 30 pounds in the first thirty days!

What is the mechanism behind this? The Atkins diet is based on the premise that the body has two fuel options: sugar from the foods you consume or stored fats in the body. Since simple sugars and carbs are the simplest to digest, the body will use them first. When you avoid consuming sugars and carbohydrates that are converted to sugars in the body, the body is left with just one fuel source: the fats you've already stored.

Since sugars aren't readily available, the body will begin to metabolize stored fats, especially those around the belly button, to provide energy. You also lose weight. That's what there is to it.

Net carbs are very simple to calculate. To find the Net Carbs in any food, take the total number of Carbohydrates listed on the nutrition info minus the amount of dietary fiber (both these units are generally shown in grams on a nutrition label), and this will equal the Net Carbs. Dietary fiber does not get processed as sugars in the body and has a negligible effect on the blood sugar levels in the system, so they are not necessary to cut.

Now that you understand Net Carbs, here are the four phases of the Atkins diet system.

Phase one allows only 20-25 grams of Net Carbs per day and does not allow any type of carbohydrate-based foods, getting the Net Carbs you eat mostly from the vegetables in your diet. This is also often called the Induction phase, meaning that your body is being inducted into eating low carbs. This is the most restrictive phase of the diet and can sometimes be tough on people who are not used o eating so few carbs.

Phase 2 can be started at least two weeks after phase 1, although most people wait until they are within 15 pounds of their goals weight. That means if you have a lot to lose, you won't move to phase 2 for a long time. Some people do choose to move to phase 2 earlier, but the weight will come off more slowly. You will still see weight loss, however. Phase 2 allows 25-50 grams of Net Carbs per day and adds foods such as whole grains and high carb fruits.

In Phase 3, you are allowed to eat from 50-80 grams of Net Carbs. You can start this phase when you are within 5 pounds of your goal weight.

Phase 4 is also called the lifetime maintenance phase, which is where you can go from 80-100 grams of Net Carbs. This is the phase that you should eat at for the rest of your life to maintain your weight loss.

After all, you want to find a happy medium between being able to eat foods that you enjoy and keeping weight off. Atkins will help you progress from very restricted Net Carbs to finding a balance for your body to have healthy carbs added while still maintaining your weight loss.

Breakfast

Farmhouse Beans & Sausage

Preparation time: 15 minutes

Cooking Time: 30 minutes

Servings: 2

Ingredients:

- 2 - cups gluten-free chicken broth
- 2 16 - oz. frozen green beans
- 1 16 - oz. chicken sausage, sliced
- ½ - onion, diced
- 2 - teaspoons Herb mare
- salt & pepper to taste

Directions:

1. Spot all fixings in the Instant Pot. Spot top on and close, ensuring the steam vent is shut.
2. Utilize manual setting and set at 6 minutes.
3. When cook time is done, utilize the fast discharge strategy to let off steam.

Nutrition:

Calories 297.6g,

Fat 8.2g,

Carbs 38.2g,

Sugars 4.8g,

Protein 19.4g

Cheesy Bacon Chicken

Preparation time: 10 minutes

Cooking Time: 40 minutes

Serving: 2

Ingredients:

- 1 2/3 to 6 - chicken breasts, cut in half widthwise
- 2/3 - tbsp. seasoning rub
- 1/8 - pound bacon, cut strips in half
- 1 1/3 - oz. shredded cheddar
- sugar-free barbecue sauce, optional, to serve

Directions:

1. Preheat stove to 400. Splash a massive rimmed preparing sheet with a cooking shower.
2. Rub the two sides of chicken bosoms with flavoring rub. Top each with a bit of bacon. Prepare for 30 min on the best rack until the chicken is 160 degrees, and the bacon looks firm.
3. Take the plate out from the broiler and sprinkle the cheddar over the bacon. Set back in the broiler for around 10 min until the cheddar is bubbly and brilliant. Present with grill sauce.

Nutrition:

Calories 345g, Fat 23g, Carbs 1g, Sugar 4.1g, Protein 29g

Baked Pesto Chicken

Preparation time: 20 minutes

Cooking Time: 40 minutes

Serving: 2

Ingredients

- 2 - chicken breasts about 1.5 lb., sliced in half widthwise to make eight pieces
- 1 1/2 - tbsp. basil pesto
- 4 - oz. mozzarella thinly sliced or shredded
- 1/4 - tsp. salt,
- 1/4 - tsp. black pepper

Directions:

1. Preheat stove to 350.
2. Splash making a ready dish with cooking shower. Spot chicken inside the base in a solitary layer and sprinkle with the salt and pepper. Spread the pesto on the bird. Put the mozzarella to finish everything.
3. Prepare for 35-forty five minutes till the chook is a hundred and sixty stages, and the cheddar is excellent and bubbly. You can hear it for a few minutes toward the conclusion to darker the cheddar inside the event you want.

Nutrition:

Calories 287.5g, Fat 16g, Carbs 4.3g, Sugars 1.8g, Protein 30.3g

Keto Cauliflower Au Gratin

Preparation time: 25 minutes

Cooking Time: 55 minutes

Serving: 2

Ingredients:

- 1/2 - large head of cauliflower, trimmed and cut into florets
- 1 - small red onions, thinly sliced
- 1 - tablespoons olive oil
- 1 - tablespoons balsamic vinegar
- 1/2 - tablespoon granular erythritol
- 1/3 - cup heavy cream
- 1/4 - cup finely grated parmesan
- 1/4 - teaspoon sea salt, more to taste
- 1/8 - teaspoon black pepper, more to taste
- 1/2 - cup shredded gruyere or gouda cheese

Directions:

1. Preheat the broiler to 350 ° F.
2. Put the olive oil in a significant skillet over medium warmth. When the dish is hot, upload the onions to the field and cook dinner till delicate and sensitive, around 15mins.
3. Include the balsamic vinegar and erythritol if utilizing, to the field with the onions, and blend to consolidate. Cook for a further 5 minutes.
4. Upload cauliflower florets to the pan and blend it with the onions.
5. In a touch bowl, is part of the large cream, parmesan cheddar, salt, and pepper? Pour the aggregate over great of the cauliflower.
6. Sprinkle the gruyere over high-quality and prepare for 40 minutes or until the cauliflower is touchy, and the splendid is first-rate dark-colored and bubbly.

Nutrition:

Calories 230g, Fat 17g, Carbs 11.9g, Sugars 0.4g, Protein 9.2g

Boosted Keto Coffee

Preparation time: 15 minutes

Cooking Time: 30 minutes

Serving: 1

Ingredients:

- 8 - ounces dark roast coffee
- 1 - tablespoons butter flavored coconut oil
- 1 - scoop Keto Zone French Vanilla
- 1 - scoop Collagen Peptides
- 2 - teaspoons monk fruit sweetened caramel syrup
- Splash coconut milk

Directions:

1. Pour all the ingredients into a blender or milk frothier and Blend till smooth and creamy.
2. ENJOY!!

Nutrition:

Calories 296g, Fat 23g, Carbs 16g, Sugars 2.1g, Protein 15g

FRESHLY BREWED
COFFEE
The perfect blend

Sugar-Free, Low Carb Dried Cranberries

Preparation time: 35 minutes

Cooking Time: 3hrs. 10 minutes

Serving: 1

INGREDIENTS

- 2 to 12 - ounce bags fresh cranberries
- 1 - Cup granular erythritol
- 3 - Tablespoons avocado oil
- ½ - teaspoon pure orange extract

Directions:

1. Preheat the stove to 200 ° F. Line two rimmed preparing sheets with material paper.
2. Wash and dry the cranberries and expel any sautéed or delicate berries. Cut the cranberries into equal parts and add them to a blending bowl.
3. Include the sugar, avocado oil, and orange concentrate if utilizing. Hurl to coat the majority of the berries equitably.
4. Line the berries in single layers over the heating sheets.
5. Heat for 3 to 4 hours, turning the racks part of the way through.

Nutrition:

Calories 208.8g, Fat 22.5g, Carbs 0.8g, Sugars 0.1g, Protein 1.6g

Original Almond Porridge

Preparation time: 5 minutes

Cooking Time: 2 minutes

Serving: 2

Ingredients:

- 1/4 cup unsweetened almond milk
- 1/3 cup ground almond
- 1/4 tsp. cinnamon
- 1-Tbsp. butter

Topping:

- 2 Tbsp. sliced roasted almond

Directions:

1. Pour the almond milk into a microwave-safe bowl. Microwave it for 2 minutes.
2. Add the almond flour to the bowl and microwave again for a minute.
3. When it is done, remove from the microwave, and then quickly stir in butter and cinnamon. Mix well.
4. Serve and enjoy warm.

Nutrition:

Each serving contains 144 Calories, 4.8 g Net Carbs, 13.1 g Fats, 4.2 g Protein

Spinach Scrambled Egg

Preparation time: 5 minutes

Cooking Time: 10 minutes

Serving: 2

Ingredients:

- 1/2 tsp. coconut oil
- two organic eggs
- 1-cup spinach
- 1/2 Tbsp. Chopped onion ring
- 1/4 tsp. pepper

Directions:

1. Crack the eggs and drop into a bowl. Using a fork, beat until incorporated, then sets aside.
2. Preheat a saucepan over medium heat, and then pour coconut oil into the saucepan.
3. Once it is hot, stir in the chopped onion and sautés until wilted and aromatic.
4. Add spinach to the saucepan and cooks until wilted.
5. Pour beaten egg over the spinach, and then quickly stir the eggs until becoming scrambles.
6. Season with pepper, then cooks until the egg is set but still soft.
7. Transfer to a serving dish, then serves immediately.

Nutrition:

Each serving contains 76 Calories, 1g Net Carbs, 5.5 g Fats, 5.8 g Protein

LUNCH RECIPES

White Lasagna Stuffed Peppers

Preparation Time: 10 minutes

Cooking Time: 30 minutes

Servings: 5

INGREDIENTS

- 2 - Large sweet peppers halved and seeded
- 1 - Tsp garlic salt divided
- 12 - Oz ground turkey
- ¾ - cup ricotta cheese
- 1 - Cup mozzarella
- 8 - Cherry tomatoes

DIRECTIONS

1. Preheat stove to 400.
2. Put the split peppers in a preparing dish. Sprinkle with 1/4 tsp garlic salt. Partition the ground turkey between the peppers and press into the bottoms. Sprinkle with another 1/4 tsp garlic salt. Heat for 30 minutes.
3. Gap the ricotta cheddar between the peppers. Sprinkle with the rest of the 1/2 tsp garlic salt. Sprinkle the mozzarella to finish everything. Put the cherry tomatoes in the middle of the peppers, if utilizing.
4. Prepare for an extra 30 minutes until the peppers are mollified, the meat is cooked, and the cheddar is brilliant.

NUTRITION: 387 calories 52g Protein 1gCarbohydrates 13g Fat

Creamy Basil Baked Sausage

Preparation Time: 3 minute

Cooking Time: 30 minutes

Servings: 4

INGREDIENTS

- 3 - Lb. Italian sausage chicken, turkey, or pork
- 8 - Oz cream cheese
- ¼ - cup basil pesto
- ¼ - cup heavy cream
- 8 - Oz mozzarella

DIRECTIONS

1. Preheat broiler to 400. Shower a huge meal dish with the cooking splash. Put the frankfurter in the heating dish. Prepare for 30 minutes.
2. In the meantime, blend together the cream cheddar, pesto, and overwhelming cream.
3. Spread the sauce over the frankfurter. Top with mozzarella. Prepare for an extra 10 minutes or until the hotdog is 160 degrees when checked with a meat thermometer.
4. Discretionary: Broil for 3 minutes to toast the cheddar to finish everything. Watch it always. It can consume effectively.

NUTRITION: 330 calorie 26g Protein 3gCarbohydrates 25g Fat

Low Carb Pork Medallions

Preparation Time: 5 minutes

Cooking Time: 30 minutes

Servings: 6

INGREDIENTS

- 1 - lb. pork tenderloin
- 3 - Medium shallots (chopped nice)
- ¼ - cup oil

DIRECTIONS

1. Reduce the beef into half inch thick cuts.
2. Hack the shallots and notice them on a plate.
3. Warm the oil in a skillet
4. Press every little bit of red meat into the shallots on the 2 aspects. The shallots will adhere to the pork in the event which you press solidly.
5. Spot the beef cuts with shallots into the warm oil and prepare dinner till accomplished. You will locate that a part of the shallots will consume for the duration of cooking, but, they'll even now confer delicious taste to the beef. Panic do not as properly. Simply cook dinner the red meat until it's cooked through.
6. Present with veggies.

NUTRITION: Carbs: 18g Fat: 37g Protein: 33g Calories: 535

Easy Mozzarella & Pesto Chicken Casserole

Preparation Time: 5 minutes

Cooking Time: 30 minutes

Servings: 4

INGREDIENTS

- ¼ - cup pesto
- 8 - Oz cream cheese softened
- ¼ - ½ - cup heavy cream
- 8 - Oz mozzarella cubed
- 2 - Lb. cooked cubed chicken breasts
- 8 - Oz mozzarella shredded

DIRECTIONS

1. Preheat stove to 400. Splash a vast meal dish with cooking shower.
2. Consolidate the initial three fixings and blend until smooth in an extensive bowl. Include the chicken and cubed mozzarella. Exchange to the goulash dish. Sprinkle the destroyed mozzarella to finish everything.
3. Prepare for 25-30 minutes. Present with zoodles, spinach, or squashed cauliflower.

NUTRITION: Calories: 404 Fat 23g Net carbs: 8g Protein: 31g

Easy Taco Casserole Recipe

Preparation Time: 5 minutes

Cooking Time: 30 minutes

Servings: 4

INGREDIENTS

- 5 to 2 - lb. ground turkey or beef
- 2 - Tbsp taco seasoning
- 1 - Cup salsa
- 16 - Oz cottage cheese
- 8 - Oz shredded cheddar cheese

DIRECTIONS

1. Preheat stove to 400.
2. Blend the ground meat and taco flavoring in a huge meal dish. Mine is 11 x 1
3. Prepare for 20 minutes.
4. In the interim, combine the curds, salsa, and 1 measure of the cheddar. Put aside.
5. Take off the meal dish from the stove and cautiously channel the cooling fluid from the meat. Separate the meat into little pieces. A potato masher works incredible for this. Spread the curds and salsa blend over the meat. Sprinkle the rest of the cheddar to finish everything.
6. Return the meal to the stove and heat for an extra 15-20 minutes until the meat is cooked completely and the cheddar is hot and bubbly.

NUTRITION: Calories: 382 Fat: 28g Net carbs: 3g Protein: 22g

Parmesan Chicken Tenders

Preparation Time: 5 minutes

Cooking Time: 20 minutes

Servings: 5

INGREDIENTS

- 1 5- lb. bag chicken tenderloins
- ¾ - cup butter
- 1⅛ - cup parmesan cheese
- ¾ - tsp. garlic powder
- Salt, to taste

DIRECTIONS

1. Soften the margarine in a skillet and include the parmesan cheddar and garlic powder (and salt, if utilizing). Plunge the chicken in the blend and spot on a treat sheet. Prepare at 325 degrees F for 20-30 minutes (until the chicken is never again pink inside and the juices run clear). Don't overbake!
2. We have likewise utilized these for Sunday lunch. My mother set them up for chapel, at that point heated them on our stove's "warm" setting for about 3½ hours while we were no more. Worked incredible!

NUTRITION: Calories: 126 Fat: 15g Net carbs: .3 g

Broccoli & Cheddar Keto Bread Recipe

Preparation Time: 5 minutes

Cooking Time: 30 minutes

Servings: 4

INGREDIENTS

- 5 - Eggs beaten
- 1 - Cup shredded cheddar cheese
- ¾ - cup fresh raw broccoli florets chopped
- 3 ½ - Tbsp coconut flour
- 2 - Tsp baking powder
- 1 - Tsp salt

DIRECTIONS

1. Preheat broiler to 350. Shower a portion skillet with a cooking splash.
2. Blend every one of the fixings in a medium bowl. Fill the portion skillet.
3. Heat for 30-35 minutes or until puffed and brilliant. Cut and serve.
4. To Reheat Microwave or warmth in a lubed griddle.

NUTRITION:

Calories: 80 Fat: 5g Net carbs: 4g Protein: 4g

Bacon Wrapped Chicken Tenders With Ranch Dip

Preparation Time: 5 minutes

Cooking Time: 45 minutes

Servings: 3

INGREDIENTS

- 12- Chicken tenderloins about 5 lb.
- 12 - Slices of bacon
- Ranch Dip Ingredient:
- 1/3 - cup sour cream
- 1/3 - cup mayo
- 1 - Tsp each garlic powder onion powder, parsley, and dill
- ½ - tsp salt

DIRECTIONS

1. Preheat the broiler to 400.
2. Wrap every chicken delicate firmly in a bit of bacon. I extended the bacon as I folded it over the chicken.
3. Spot on a heating sheet. Prepare for 35-45 minutes until the bacon is fresh and the chicken is completely cooked.
4. In the meantime, mix together the elements for the plunge. Present with the cooked chicken.

NUTRITION:

Calories: 473 Fat: 40g Net carbs: 6g Protein: 21g

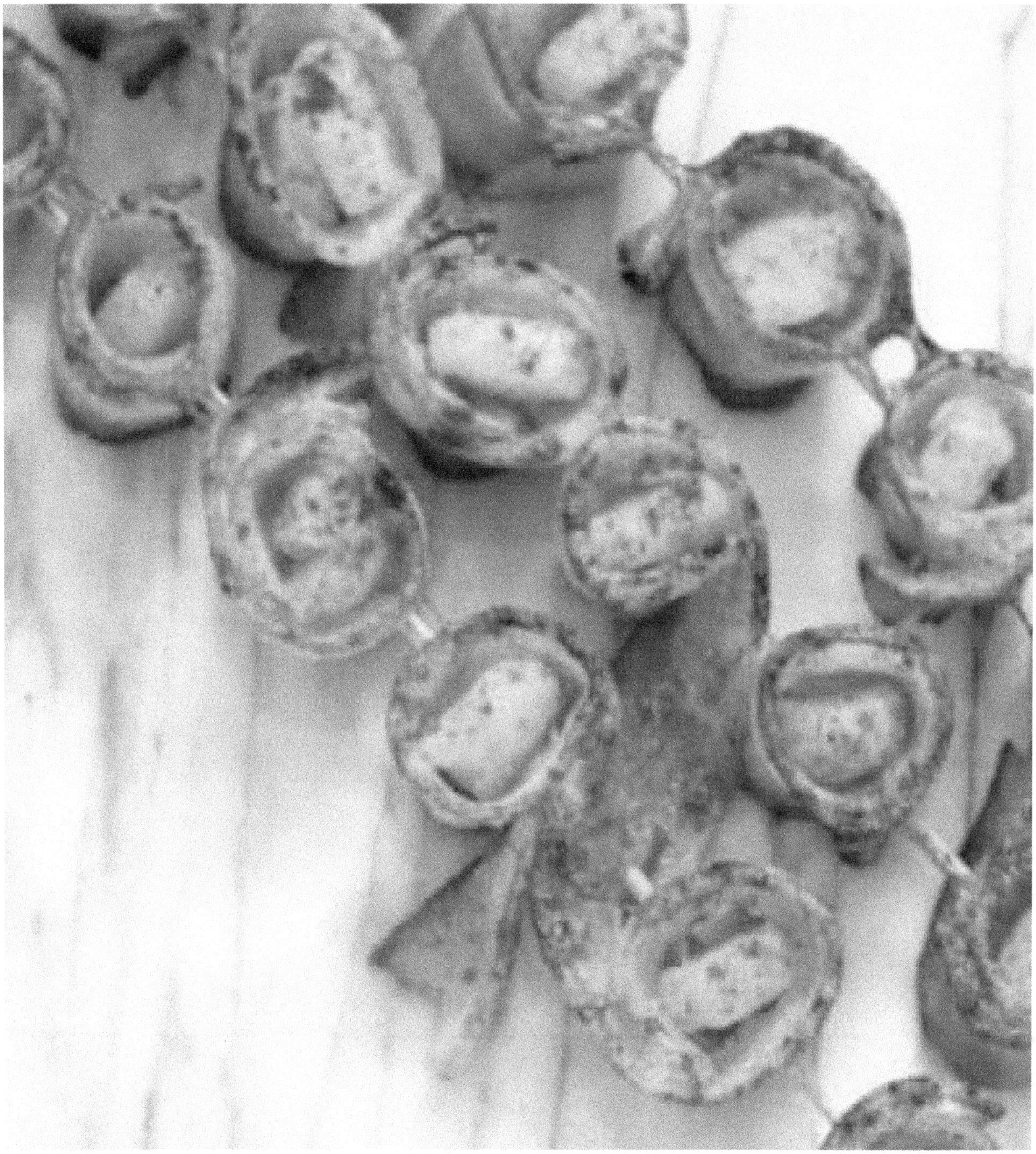

DINNER RECIPES

Basic Pork Shanks

Preparation time: 10 minutes

Cooking Time: 20 minutes

Serving: 2

Ingredients

- 2 pork shanks
- 1½ teaspoons salt
- ½ teaspoon black pepper
- 2 garlic cloves, crushed
- 2 tablespoons olive oil
- 5 to 6 sprigs fresh rosemary
- 2 cups beef broth, divided
- 2 tablespoons balsamic vinegar

Directions:

1 Place the pork shanks in a baking pan.

2 Whisk rest of the ingredients in a bowl and pour over the shanks.

3 Place these shanks in the Air fryer basket.

4 Press "Power Button" of Air Fry Oven and turn the dial to select the "Air Fry" mode.

5 Press the Time button and again turn the dial to set the cooking time to 20 minutes.

6 Now push the Temp button and rotate the dial to set the temperature at 360 degrees F.

7 Once preheated, place the Air fryer basket in the oven and close its lid.

8 Slice and serve warm.

Nutrition:

Calories 388

Total Fat 9.7 g

Saturated Fat 4.7 g

Cholesterol 181 mg

Sodium 245 mg

Total Carbs 12.5 g

Fiber 0.3 g

Sugar 1.8 g

Protein 30.3 g

Easy Italian Zucchini Meatloaf

Preparation time: 10 Minutes

Cooking time: 35 Minutes

Serving: 2

Ingredients

- 2 lbs. extra lean ground beef
- 2 large eggs
- 1 cup shredded zucchini
- 1/2 cup grated Parmesan cheese
- 1/2 cup finely chopped parsley
- 2 crushed garlic cloves
- 2 tbsp. balsamic vinegar
- 1 tbsp. dry oregano
- 2 tbsp. minced dry onion or onion powder
- Salt to taste
- 1/2 tsp ground black pepper
- Cooking coconut oil spray

Directions:

1 Place all the ingredients in a large bowl and mix thoroughly.

2 Prepare the slow cooker by lining with aluminum foil with flaps on the outside, to make it easier to remove the meatloaf later. Spray the lining with coconut oil spray.

3 Put the mixture in cooker. Cover and cook on Low for about 6 hours, or on High for 2-2.5 hours.

Nutrition: Protein: 20.7g, Fat: 9 g, Carbohydrate: 3.4 g, Fiber: 0.1 g Total Calories: 97.5

Healthy Green Soup

Preparation time: 15 Minutes

Cooking time: 15 Minutes

Serving: 2

Ingredients:

- 1 medium-sized cauliflower
- 1 medium-sized broccoli
- 2 leeks
- zucchini
- 1 onion
- 1 cup of water
- 2 garlic cloves
- ½ cup of butter
- Salt
- ¼ cup of collagen protein
- Optional Seasonings (basil, oregano, rosemary, thyme, etc.)

Directions:

1. Wash all the vegetables and remove any damaged parts or fibrous pieces.
2. Chop roughly all of the vegetables and place them in a large soup pan.
3. Add in the remaining ingredients.
4. Cook over medium warmth until the vegetables are tender.
5. To achieve a thicker soup, allow the soup mixture to simmer longer.
6. When completed, pour the soup mixture into your blender or food processor.
7. Puree until creamy smooth.
8. Add salt as desired and serve.

Nutrition:

Calories 425

Fats 34 g

Protein 12 g

Carbohydrates 16 g

Fiber 3.6 g

Cheddar Cream of Broccoli Soup

Preparation time: 15 Minutes

Cooking time: 15 Minutes

Serving: 2

Ingredients:

- 1 onion
- 2 minced garlic cloves
- 2 tablespoons of unsalted butter
- 2 tablespoons of almond flour
- 2 cups of heavy cream
- 2 cups of vegetable broth
- 7 to 10 cups of broccoli florets
- 1.5 cups of grated sharp white cheddar cheese
- A Pinch of nutmeg
- Salt
- Pepper
- Optional sour cream or Greek yogurt for garnishing

Directions:

1 Dice the onion.
2 In a soup pot, melt the unsalted butter over medium heat.
3 Add in the onion and cook for five minutes until tender.
4 Add in the minced garlic and sauté for about a minute.
5 Now add in the flour and cook for three minutes, continually stirring until it is golden.
6 Add in the cream and vegetable broth, raising the heat, and bring the mixture to a boil.
7 Go the heat to medium and add in the broccoli. Simmer for four to five minutes.
8 Remove four broccoli florets and wash them in cold water. Set them aside.
9 Decrease the heat to a low situation and simmer partially covered for few minutes until the broccoli is tender.
10 Puree the soup in your blender or by using an immersion blender. Pour the soup mixture into a soup-serving bowl.
11 Add the cheddar, nutmeg and salt and pepper, and whisk.
12 Prepare the soup in individual bowls, garnished with yogurt or sour cream and the cooked broccoli floret previously set aside.

Nutrition:

Calories 588

Fat 48.8 g

Protein 21.6 g

Carbohydrates 20.5 g

Fiber 6.5 g

Delightful Garden Salsa

Preparation Time: 5 minutes

Cooking time: 10 minutes

Serving: 2

Ingredients:

- 2 cups tomatoes, peeled and diced
- 1 onion, chopped
- 2 tablespoons coconut oil, melted
- 2 large green bell peppers, chopped
- 2 jalapeno peppers, chopped
- ½ cup chili pepper, minced
- 3 garlic cloves, minced
- 1 tablespoon cumin
- 2 tablespoons salt
- 1 tablespoon cayenne pepper sauce
- 1 teaspoon garlic powder
- ½ cup apple cider vinegar
- 1 (15-ounce) can tomato sauce
- 1 (6-ounce) can tomato paste

Directions:

1 Place all the ingredients in the Instant Pot.

2 Close and seal the lid; cook at high pressure for 10 minutes.

3 When the cooking is done, naturally release the pressure and remove the lid.

4 Place the salsa in your refrigerator and allow to chill.

5 Serve and enjoy with keto tortilla chips.

Nutrition:

Energy (calories): Total: 655 kcal Per one serving: 140 kcal

Calorie Breakdown:

Protein: 3.5g Fat: 10%

Carbohydrates: 4%

Carbohydrates mass fraction: 1%

Chicken Meatballs

Preparation Time: 10 minutes

Cooking Time: 12 Minutes

Serving: 2

Ingredients:

- 1 lb. ground chicken
- 1/3 cup frozen spinach, drained and thawed
- 1/3 cup feta cheese, crumbled
- 1 tsp Greek seasoning
- ½ oz pork rinds, crushed
- Pepper
- Salt

Directions:

1. Spray air fryer basket with cooking spray.
2. Add all ingredients into the large bowl and mix until well combined.
3. Make small balls from meat mixture and place into the air fryer basket and cook for 12 minutes.
4. Serve and enjoy.

Nutrition:

Calories 271

Fat 12 g

Carbohydrates 1 g

Sugar 0.5 g

Protein 37 g

Cholesterol 117 mg

Spinach Breakfast Parcels

Preparation time: 10 minutes

Cooking time: 4 minutes

Servings: 2

Ingredients:

- 1 sheets filo pastry
- 1 pound baby spinach leaves, roughly chopped
- ½ pound ricotta cheese
- 2 tablespoons pine nuts
- 1 eggs, whisked
- Zest from 1 lemon, grated
- Greek yogurt for serving
- Salt and black pepper to the taste

Directions:

1 In a bowl, mix spinach with cheese, egg, lemon zest, salt, pepper and pine nuts and stir.

2 Arrange filo sheets on a working surface, divide spinach mix, fold diagonally to shape your parcels and place them in your preheated air fryer at 400 degrees F.

3 Bake parcels for 4 minutes, divide them on plates and serve them with Greek yogurt on the side.

4 Enjoy!

Nutrition: calories 182, fat 4, fiber 8, carbs 9, protein 5

Ham Rolls

Preparation time: 10 minutes

Cooking time: 10 minutes

Serving: 2

Ingredients:

- 1 sheet puff pastry
- 1 handful gruyere cheese, grated
- 1 teaspoons mustard
- 2 ham slices, chopped

Directions:

1 Roll out puff pastry on a working surface, divide cheese, ham and mustard, roll tight and cut into medium rounds.

2 Place all rolls in air fryer and cook for 10 minutes at 370 degrees F.

3 Divide rolls on plates and serve for breakfast.

4 Enjoy!

Nutrition: calories 182, fat 4, fiber 7, carbs 9, protein 8

Snacks

Sheet Pan Shrimp Fajitas Recipe

Preparation Time: 10 minutes

Cooking Time: 15 minutes

Servings: 4

INGREDIENTS

- 1 red bell pepper, seeded and julienned.
- 1/2 green bell pepper, seeded and julienned.
- 1/2 big yellow onion, very finely sliced.
- 2 teaspoons plus 1/3 cup canola oil, divided. 2garlic cloves.
- 1 teaspoon kosher salt.
- 1/2 teaspoon dried oregano.
- 1/2 teaspoon chili powder.
- 1/2 teaspoon sweet paprika.
- 1/2 teaspoon cayenne pepper.
- 1/4 teaspoon ground cumin.
- 1 pound (41/50 size) raw shrimp, deveined and shelled.
- 6 flour tortillas.
- Mexican crema, sour cream, or Greek yogurt.

DIRECTIONS

1. Preheat the oven to 400 ° F. Set the sheet pan nearby.
2. Juice 1 1/2 of the limes. Cut the other lime half into 6 wedges and set them aside.
3. Toss the bell peppers and onion into the 2 teaspoons of canola oil till layered. Spread them onto the sheet pan in a single layer.
4. Put the lime juice and garlic into a mixer, including in the staying oil in a stable stream. Marinade the shrimp, tossing to coat in a big zip-seal bag for 15 minutes.
5. Roast the bell pepper and onion for 10 minutes.
6. Get rid of the shrimp from the marinade. Polka dot the shrimp onto the sheet pan of roasted veggies.

7. Heat some flour tortillas and set out the Mexican crema or sour cream to serve with the lime wedges.

NUTRITION: Calories: 26 Fat: 17g Net carbs: 6g Protein: 20g

Macadamia Nut Alfredo Zoodles

Preparation Time: 15 minutes

Cooking Time: 25 minutes

Servings: 4

INGREDIENTS

- 1 cup roasted macadamia nuts.
- 1 cup warm water, plus more as required.
- 1/2 cup dietary yeast.
- Passion and juice of 1/2 lemon, or more to taste.
- 1/2 teaspoon garlic salt.
- 4 tablespoons additional virgin olive oil, divided.
- 4 zucchini, ends cut off.
- Parsley, sliced, to garnish.
- 1/4 cup roasted macadamia nuts, sliced, to garnish (optional).
- Newly ground black pepper, to decorate.

DIRECTIONS

1. In a high-speed mixer, include macadamia nuts and hot water. When softened include dietary yeast, lemon enthusiasm and juice, garlic salt, 2 tablespoons of olive oil. Taste and change spices if required.
2. Spiralize zucchini onto a cutting board or bowl.
3. In a sauté pan, heat 2 tablespoons of oil over medium-high heat. Include spiralized zucchini.
4. Prepare till intense and wilted green (a couple of minutes), seasoning with garlic salt.
5. Gather sauce. Toss to coat, and permit it to heat through for a couple of minutes.
6. Transfer zoodles to plates.
7. Garnish with parsley, sliced macadamia nuts, and black pepper.

NUTRITION: Calories: 208 Total Fat: 14g Saturated Fat: 4g Cholesterol: 16mg Sodium: 472mg Carbohydrates: 15g Fiber: 6g Added Sugars: 0g; Protein: 9g Potassium: 591mg Vitamin K: 201mcg

Greek Yoghurt With Honey & Walnuts

Preparation Time: 1 minute

Cooking Time: 1 minute

Servings: 1 1x

INGREDIENTS

- 3/4 cup 0%
- Fat Greek Yoghurt
- 4 halves walnuts
- 1/4 tsp ground cinnamon
- 1 tsp Greek Honey

DIRECTIONS

1. Put all the active ingredients in a charming dessert bowl.
2. Stir well and delight in.

NUTRITION: Calories: 84 Fat: 7g Net carbs: 6g Protein: 3g

Chocolate Peanut Butter Cake

Preparation Time: 35 minutes. + Cooling

Cooking Time: 25 minutes. + Cooling

Servings: 4-6

INGREDIENTS

- 2 cups sugar
- 1 cup 2% milk
- 1 cup strong brewed coffee
- 1 cup canola oil
- 2 big eggs, space temperature level
- 1 teaspoon vanilla extract
- 2 cups versatile flour
- 3/4 cup baking cocoa
- 1 tablespoon immediate coffee granules
- 2 teaspoons baking soda
- 1 teaspoon salt

Filling:

- 1 cup butter, softened
- 3/4 cup velvety peanut butter
- 1-1/2 cups confectioners' sugar

Icing:

- 1/2 cup velvety peanut butter
- 6 ounces bittersweet chocolate, sliced
- 1 cup butter, softened
- 2 cups marshmallow creme
- 1/3 cup confectioners' sugar
- 3/4 teaspoon vanilla extract
- 2 cups dark chocolate chips, optional
- 1 cup peanut butter chips, optional
- 1 teaspoon reducing, optional

DIRECTIONS

1. In a big bowl, beat the sugar, milk, coffee, oil, eggs and vanilla up until well combined. Integrate the flour, cocoa, coffee granules, and baking soda and salt; slowly beat into sugar mix till combined.
2. Transfer to 2 floured and greased 9-in. round baking pans. Bake at 350 ° for 25-30 minutes or till a toothpick placed in the center comes out tidy.
3. Cool for 10 minutes before getting rid of from pans to wire racks to cool totally.

4. For the filling, in a big bowl, beat butter and peanut butter till mixed. Include confectioners' sugar; beat up until smooth.
5. For the icing, melt peanut butter and chocolate in a microwave; stir up until smooth. In a big bowl, beat butter and chocolate mix up until fluffy.
6. Location 1 cake layer on a serving plate; spread out with filling. Chill for at least 1 hour before serving.
7. For optional garnish, location chocolate chips in a microwave-safe bowl; microwave 1-2 minutes or till melted, stirring every 20 seconds. Repeat with peanut butter chips and reducing. Drizzle melted peanut butter mixture over the melted chocolate and swirl with the suggestion of a spoon.

NUTRITION: 1 piece (computed without chocolate curls): 883 calories 60g Fat (23g hydrogenated Fat) 100mg cholcstcrol 670mg salt 83g carb (60g sugars, 4g fibcr) 11g Protcin.

Low Carb Cupcakes

Preparation Time: 15 minutes

Cooking Time: 25 minutes

Servings: 5

INGREDIENTS

- 1/3 cup coconut flour.
- 1/2 cup unsweetened cocoa powder.
- 1/4 cup powdered erythritol (another low carbohydrate sweetener of your option will work).
- 1 teaspoon baking powder.
- 1/2 teaspoon baking soda.
- 1/4 teaspoon salt.
- 4 entire eggs.
- 1 teaspoon vanilla extract.
- 8 drops stevia extract optional for new sweet taste.
- 4 Tablespoons additional light olive oil.
- 1/2 cup unsweetened almond milk (or another dairy-free option).

DIRECTIONS

1. Preheat oven to 350 degrees F. Prepare a muffin tin by baking or greasing with cupcake liners.
2. In a medium bowl blend coconut flour, cocoa powder, erythritol, baking powder, baking soda, and salt.
3. Include eggs, vanilla extract, stevia (if including), olive oil and almond milk. Mix till active ingredients are well integrated, if mix ends up being thicker in consistency than you 'd like to feel complimentary to include 2 Tablespoons of water to damage till it reaches your desired consistency.
4. Spoon 2 Tablespoons of batter into each tin. Bake 20-22 minutes, or till toothpick comes out tidy.

NUTRITION:

Calories: 177 Total Fat: 7g Saturated Fat: 1g Cholesterol: 0mg Sodium: 192mg Carbohydrates: 27g Fiber: 4g Added Sugars: 0g; Protein: 2g Potassium: 454mg Vitamin K: 8mcg

Low Carb Vegan Coconut Matcha Cupcakes

Preparation Time: 10 minutes

Cooking Time: 22 minutes

Servings: 12 cupcakes.

INGREDIENTS

- 1/4 cup, plus 1 tablespoon (so, 5 tablespoons) coconut flour
- 1/2 cup coconut butter/manna
- 1 tablespoon ground flax seeds
- 1/4 cup Swerve, or other granulated sweeteners
- 1/2 cup warm water
- 1 tsp vanilla extract
- 1 tsp baking powder
- Pinch of salt

For the Frosting:

- 1 cup raw cashews
- 1/2 cup full-fat coconut milk (from a can!).
- 2 tablespoon Swerve, or other granulated sweeteners.
- 1 tsp vanilla extract.
- 1/2 tsp matcha powder.

DIRECTIONS

To Make the Cupcakes.

1. Preheat oven to 350F. Grease or line 6 wells of a muffin tin (or utilize a silicone one!).
2. Put water over coconut butter, and stir up until combined and smooth.
3. Mix in flax, vanilla, Swerve and salt. Let this sit for a couple of minutes for the flaxseed to get all.
4. In a different bowl, integrate baking powder and coconut flour.
5. Gradually include flour mix, and stir up until there are no swellings and whatever is even.
6. Divide into muffin tins and bake for 20-25 minutes, till the tops are firm and the edges are only beginning to turn golden.

7. Get rid of the oven and either wait a couple of hours for these to cool, or pop them in the freezer for 30 minutes while you make the icing. These must set up entirely before being gotten rid of, or they will fall apart.

NUTRITION: 223 calories 21gCarbohydrates 7gProtein Fiber6g

Flourless Chocolate Cake

Preparation Time: 15 minutes

Cooking Time: 45 minutes

Servings: 12

INGREDIENTS

- 1
- /3 cup water
- 1/4 teaspoon salt
- 1/2 cup low carbohydrate sugar replacement
- 12 ounces unsweetened baking chocolate
- 2/3 cup butter or ghee, cut into tablespoon-size pieces
- 4 big eggs
- Boiling water

DIRECTIONS

1. Line bottom of 9-inch springform pan with parchment paper.
2. In little pot, heat water, salt, and Swerve over medium heat up until salt and sweetener are liquefied.
3. Melt baking chocolate in a double boiler or microwave.
4. Mix melted chocolate and butter in a big bowl with electrical mixer.
5. Beat in the warm water mix.
6. Include egg, one at a time, beating well after including each.
7. Put mix into ready spring form pan. Wrap outside well with foil.
8. Location spring form pan in a bigger cake pan and include boiling water to the outdoors pan about 1 inch deep.
9. Bake cake in a water bath for 45 minutes at 350 ° F. Remove and cool a little on a cake rack.
10. Chill cake overnight in the fridge. Eliminate side of springform pan.

NUTRITION: Calories: 214 Total Fat: 7g Saturated Fat: 3g Cholesterol: 11mg Sodium: 168mg Carbohydrates:32gFiber: 4g Added Sugars: 0g;Protein: 7g Potassium: 801mg Vitamin K: 19mcg

Low Carbohydrate Yum

Preparation Time: 20 Minutes

Cooking Time: 15 minutes

Servings: 24

INGREDIENTS

- 144 g almond flour
- 37 g cocoa powder
- 13 g black cocoa powder or just more routine cocoa
- 3/4 teaspoon kosher salt
- 1/2 teaspoon xanthan gum
- 1/2 teaspoon baking soda
- 1/4 teaspoon espresso powder or instant coffee (optional).
- 80 g saltless grass-fed butter at space temperature level.
- 128 g erythritol.
- 1 egg.
- For Vanilla Cream Filling. 56 g saltless grass-fed butter.
- 14 g coconut oil.
- 1 1/2 teaspoon vanilla extract.
- Pinch kosher salt.

DIRECTIONS

1. 63-125 g powdered erythritol or powdered sweetener, to taste (we utilize.
2. Present the dough in between 2 pieces of parchment paper up until thin and fabulous. Cutout the rounds (Oreos are approximately 1 3/4 inches in size).
3. Similar to the initial, anticipate from these keto Oreo cookies a tasty vanilla cream sandwiched in between 2 (appropriately crisp!) dark chocolate cookies ... however at only 1g net carbohydrate a pop!
4. With your mixer on low, include half of your flour mix- blending up until simply covered. Mix in the rest.
5. Wrap cookie dough with stick movie (saran wrap) and cool for 1 hour (or overnight).
6. Start to cream butter in a big bowl with an electric mixer, 1-2 minutes. Include sweetener and continue to beat till much and completely blended of the sweetener has liquified (3-5 minutes).
7. To make the vanilla cream filling, cream butter and coconut oil (or more butter) in a medium bowl with an electric mixer. Include vanilla extract and a pinch of salt, and mix till included.

Include powdered sweetener to blend and taste till totally integrated and fluffy and light in texture.

Dish Notes.

8. Adding coconut oil to the vanilla cream enhances the texture a fair bit. It likewise includes little taste; however, you can continuously utilize refined coconut oil (unappetizing) or just add more butter.
9. Oh, and if baking with cups instead of grams is your thing, simply click United States Cups for an instantaneous conversion.

NUTRITION: Fat 8g HydrogenatedFat3g Cholesterol 19mg Salt 138mg6%. Potassium 34mg1%. Carbohydrates2g1%. Fiber 1g4%. Protein 1g2%. Vitamin A 150 3%. Calcium 18mg2%. Iron 0.5mg

DESSERTS

Blackberry Yogurt Nuts (Phase 2)

Preparation time: 10 minutes

Cooking Time: 1hr minutes

Serving: 2

Ingredients:

- 1 cup plain, unsweetened Greek yogurt
- Macadamia nuts
- 1/2 cup fresh blackberries

Directions

1. In a food processor, combine yogurt and nuts until paste forms.
2. Combine pasta and blackberries in a small bowl, mixing them gently until combined.
3. Lay coated blackberries out on a baking sheet in a single layer and refrigerate for 1 hour.

Nutrition:

Calories 43

Total Carbohydrate 10 g

Dietary fiber 5 g

Sugar 4.9 g

Protein 1.4 g

Cherry Plum Bread Pudding (Phase 3)

Preparation time: 15 minutes

Cooking Time: 45 minutes

Serving: 2

Ingredients:

- 1 TBS. butter
- slices whole-wheat bread, stale
- Three ripe plums, chopped
- 1 cup pitted cherries
- cups unflavored almond milk
- Four eggs, beaten
- 1 tsp. cinnamon
- TBS. cream cheese

Directions

1 Preheat oven to 350 degrees. Grease the bottom and sides of a 9"x13" pan with butter. Arrange torn chunks of bread over the bottom of the pan.

2 In a large mixing bowl, combine fruit, milk, eggs, cinnamon, and cream cheese. Whip together until well combined.

3 Pour mixture over bread, and bake for 45 minutes, or until the top is well browned.

Nutrition:

341 calories; total fat 11g; saturated fat 5g;

Cream Cheese Energy Balls (Phase 4)

Preparation time: 2 hrs. and 20 minutes

Cooking Time: 0 minutes

Serving: 2

Ingredients:

- 1/2 cup cream cheese
- 2 Brazil nuts
- 1 cup shredded coconut
- Six fresh dates
- TBS. oat bran

Directions

1 Combine cream cheese, nuts, dates, and oat bran in a food processor. Blend until completely incorporated.

2 Form into 1" balls, roll in coconut, and place on a cold baking sheet.

3 Refrigerate for 2 hours.

Nutrition:

Calories: 109.6

Dietary Fiber: 2.4 g

Protein: 2.5 g

Sugars: 7.1 g

Classic Almond Flour Pound Cake

Preparation time: 30 minutes

Cooking Time: 50 to 55 minutes

Serving: 2

Ingredients:

- 1 cup Splenda
- 1 cup or two sticks butter, softened at room temperature
- 2 cups almond flour
- One teaspoon baking powder
- Five eggs, at room temperature
- One teaspoon vanilla extract
- One teaspoon lemon extract

Directions:

1. In a large bowl, cream one cup of Splenda and two sticks of softened butter well. After this, add five eggs one by one, beating well the mixture before adding in the egg.
2. Combine two cups of almond flour with one teaspoon of baking powder in a separate bowl.
3. Slowly mix the wet mixture into the dry one with continuous beating.
4. Add a teaspoon each of vanilla and lemon extracts. The same goes for the color of the cake; just add food color if desired.
5. Pour mixture into a greased nine-inch or ten-inch baking pan. A spring form pan or a round cake pan can be used. Bake for fifty to fifty-five minutes at 350°F.
6. Let it cool for about twenty to thirty minutes after baking. Top with cream cheese frosting if desired. Makes twelve servings.

Nutrition:

143Cal, 4gCarbs. 26gFat, 9gProtein

Flourless Chocolate Pecan Torte

Preparation time: 20 minutes

Cooking Time: 30 minutes

Serving: 2

Ingredients:

- 2 cups unsalted pecans
- ⅓ cup of cocoa powder
- ¼ cup erythritol, if desired
- ½ cup or one stick melted butter
- ½ cup of water
- 1 cup sugar equivalent of artificial sweetener of choice
- Four eggs
- One teaspoon vanilla extract
- One teaspoon baking powder

Directions:

1 Preheat oven to 350 °F.

2 Pulse two cups of unsalted pecans in a food processor until they are meal, but not as small as cornmeal.

3 Add one-third of cocoa powder, a cup of artificial sweeteners, and a teaspoon of baking powder into the food processor and pulse again. Also, add erythritol if desired.

4 After the dry ingredients, add half a cup of melted butter, another half cup of water, four eggs, and a teaspoon of vanilla extract and process until everything is combined well.

5 Grease an eight-inch or nine-inch spring form pan with cooking spray. Pour batter into the pan and bake at 350 °F for about thirty minutes, or until a toothpick comes out clean when inserted in the middle of the cake. It is best to check if the cake is made twenty-five minutes after putting it in the oven.

6 Let it cook for another ten minutes. Serve with chocolate sauce if desired.

Nutrition:

340Cal. 28gCarbs. 25gFat. 4gProtein

Chocolate Torte Cake

Preparation time: 20 minutes

Cooking Time: 25 minutes

Serving: 2

Ingredients:

For the torte:

- 1½ cups Splenda
- One teaspoon almond extract
- ½ teaspoon cream of tartar
- Six egg whites

For the chocolate cream filling:

- 1½ cups heavy cream
- ½ cup Splenda
- ¼ cup unsweetened cocoa powder
- ¼ cup Crème de Cacao
- ½ teaspoon vanilla extract
- ¼ cup slivered almonds, for topping, if desired
- Melted unsweetened chocolate or chocolate syrup, if desired

Directions:

1 Preheat oven to 300 ° F.

2 For the chocolate cream filling, mix one and a half cups of heavy cream, one-half cup of Splenda, one-fourth cup of unsweetened cocoa powder, and one-half teaspoon vanilla extract in a medium bowl. Beat until the mixture is thick, and then add Crème de Cacao. Continue beating until stiff. Set aside.

3 For the torte, mix six egg whites with one teaspoon of almond extract and half a teaspoon of cream of tartar. Beat until soft peaks form. After this, slowly add a cup and a half of Splenda. Mix well.

4 Coat cookie sheets with cooking spray (or Pam). Put meringue on the cookie sheets, forming six-inch circles on each.

5 Put inside the preheated oven and bake at 300 ° F for twenty-five minutes or until meringue turns to light brown.

6 Remove from cookie sheets immediately and let them cool for about ten to fifteen minutes.

7 On a plate, place one layer of meringue and top with a third of the chocolate cream filling. Repeat step until all meringues and chocolate cream filling are used.

8 Place inside the refrigerator overnight

9 Before serving, sprinkle slivered almonds and drizzle with melted unsweetened chocolate or chocolate syrup on top if desired. Makes ten servings.

Nutrition:

410Cal. 54gCarbs. 19gFat. Three %3gProtein

Chocolate Angel Food Cake

Preparation time: 15 minutes

Cooking Time: 20 minutes

Serving: 2

Ingredients:

- 6 to 10 packets of Splenda
- Eight eggs, separated
- ¼ cup soy flour
- ¼ teaspoon cream of tartar
- 1tbsp. unsweetened cocoa powder
- One sachet of the fat-free cocoa mix (Swiss Miss Fat-Free with Splenda recommended)
- 1tbsp.chocolate-flavored protein powder

Directions:

1 Preheat oven to 350°F.

2 Beat eight egg whites in a medium bowl, slowly adding cream of tartar one-fourth teaspoon of cream of tartar. Add three packets of Splenda while beating the eggs and cream of tartar to a stiff but not dry consistency.

3 In another medium bowl, combine eight egg yolks with Splenda's remaining packets, the number of which depends on personal sweetness level preference. Add half of the fat-free cocoa mix sachet and half a teaspoon of the unsweetened cocoa powder. Mix well.

4 Combine one-fourth cup of soy flour with a tablespoon of chocolate-flavored protein powder and the remaining cocoa mix and unsweetened cocoa powder in a third bowl.

5 Take one tablespoon of the first mixture and combine it with the second. After mixing the combinations well, pour in the second mixture into the remainder of the first. Sift the dry mix over the combined yolk and white combination and fold them nicely.

6 Coat a Bundt pan or a loaf pan with a cooking spray (or Pam). Pour the well-combined batter into the pan and bake it for twenty minutes at 350°F.

7 Let it cool for about twenty minutes. Top with peanut butter and chocolate frosting if desired.

Nutrition:

130Cal. 29gCarbs. 0 Fat. 3gProtein

Cheesecake

Preparation time: 45 minutes

Cooking Time: 1hr to 40 minutes

Serving: 2

Ingredients:

For the crust:

- Two tablespoon sugar equivalent of artificial sweetener of choice
- Two tablespoons melted butter
- 1 cup almond meal

For the filling:

- Four eggs, at room temperature
- 1½ pounds cream cheese, at room temperature
- ¼ cup sour cream
- 1⅓ cups sugar equivalent of artificial sweetener of choice
- 1½ teaspoons lemon juice
- 1½ teaspoons vanilla extract

Directions:

1 Preheat oven to 375°F.
2 In a large bowl, combine two tablespoons of artificial sweetener, two tablespoons melted butter, and one-cup almond meal to form the crust.
3 Put mixture into a spring form pan and press to the bottom. Bake for eight to ten minutes, or until lightly browned and fragrant.
4 After baking the crust, decrease oven temperature to 350°F if using a water bath for baking, or increase to 400°F if not.
5 In another large bowl, beat one and a half pounds of cream cheese until fluffy.
6 Add four eggs, one-fourth cup of sour cream, one-third cups of artificial sweetener, one and a half teaspoons of lemon juice, and one and a half teaspoons of vanilla extract. Continue beating, scraping the bowl and beater's sides every time so that ingredients are fully incorporated.
7 After ingredients are well combined, scrape the bowl's sides and the beater more time and beat the mixture for another minute.
8 Pour mixture into the pan over the crust.
9 If using a water bath (at 350°F): Using foil, wrap the sides and the pan's bottom. Put the spring form pan in a bigger baking pan. Pour water into the baking pan, ensuring that water does not spill on the spring form pan where the cheesecake is. Bake for an hour to an hour and a half. When the center reaches 155°F or when the cake is firm, the center is slightly soft, removing the cheesecake from the oven.

10 If not using a water bath (at 400 °F): Put the spring form pan on top of a sheet pan just in case the cream cheese filling drips. Put the cheesecake into the oven at 400 °F, and immediately turn down the temperature to 200 °F. Bake for an hour to an hour and a half. Check the temperature of the center after an hour. When it has reached 155 °F or when the cheesecake is firm, the middle is slightly soft, removing the oven's cheesecake. (Note: An unglazed ceramic tile, brick, or a pizza stone will help hold the heat in the range, ensuring an even temperature throughout the baking process. Put one of either three at the lower rack, if available, when not using a water bath in baking a cheesecake.)

11 Chill overnight in a refrigerator.

12 Top with chopped strawberries or raspberries if desired. Makes sixteen servings.

Nutrition:

Calories: 472 Carbs: 54g Fat: 18g Protein: 20g

7-day meal plan

Day	Breakfast	Lunch	dinner	Desserts
1	Parfait with Varieties Of Berries For Breakfast	Steak Tomahawk Grilled	Broccoli Soup, Green Leaves, and Beans	Raspberry Parfait
2	Burrito For Breakfast	Grilled Seafood Salad And Salsa Verde With Thai Basil	Enjoyable Shrimp Paella	Brownie Drops
3	Stuffed Portobello Broilers	Grilled Chicken With Ranch Sauce	Celestial Seafood Gumbo	French Quesadillas
4	Bacon, Egg & Cheese Casserole	Barbecue Peanut Chicken Skewers	Vegetarian Cauliflower Crust Pizza with Mushrooms and Olives	Chocolate Coconut Smoothie
5	Vanilla and Marshmallow Smoothie	Grilled Beef Tacos With Sweet Potatoes	Cream of pear and arugula	Mint Chocolate Smoothie
6	Pumpkin Almond Pancakes	Recipe Of Boar Shoulder On The Smoker	Baked Brie with Tomatoes and Nuts	Coffee Cake Muffins
7	Eggs With Salsa And Avocado	Smoker - Hot Smoked Trout	Cream of Garlic Chicken Soup	Atkins Almond Flour Pancakes

Conclusion

As low-carb diets have increased in popularity, a variety of studies have been performed by researchers to determine the efficacy and additional health effects of the diets. The Harvard School of Public Health states that some evidence suggests that low-carb diets may help individuals lose weight faster and manage it more than low-fat diets can and that discreetly low-carb diets can be heart-healthy as long as healthier sources are responsible for protein and fat choices.

Twenty-three randomized controlled trials of more than 2,500 participants were looked at in a previous meta-analysis. The researchers found that participants on a low-carb diet experienced a slight yet statistically important decrease in overall levels of cholesterol and triglycerides and at least comparable decreases in weight, waist circumference, and other risk factors for disease relative to participants on a low-fat diet. Another research showed that low-carb dieters were able to sustain weight loss because, compared with those adopting a diet with higher carbs, they burned about 200 additional calories per day.

The low-carb diet works like this: the body is instead forced to burn fat for fuel when carb intake is low, producing ketone bodies that are used for energy. Ketones are by-products of metabolism, and when the body turns to burn fat rather than carbohydrates, they are created during a phase called ketosis, a common occurrence in certain individuals who adopt low-carb diets, such as the keto diet and Atkins diet.

With Atkins, you don't need to work out to lose weight, but you should get going. Try exercising every day for 30 minutes or more. If you have been inactive or have any medical conditions, talk to your doctor first.

Limiting starchy, sugary carbohydrates can help reduce calories and allow for weight loss for the individual who needs structure in their diet. And the safe and smart thing to do is to concentrate on proteins and fats that are plant-based.

You have to move on from the original Atkins 20 diet for your long-term wellbeing. It's the later stages of the diet that offer you a variety of foods that are necessary for health, especially the Atkins 40. As you begin eating seeds, starchy vegetable beans, whole grains, fruits, and nuts again, you have to exercise and keep portions small.

CPSIA information can be obtained
at www.ICGtesting.com
Printed in the USA
LVHW011035010621
689024LV00025B/2401

9 781802 851588